Guilherme Dilarri
Carolina Rosai Mendes
Carlos Renato Corso

Comparative physical-chemical study in the treatment of textile effluent

Guilherme Dilarri
Carolina Rosai Mendes
Carlos Renato Corso

Comparative physical-chemical study in the treatment of textile effluent

A study with Saccharomyces cerevisiae yeast and chitosan powder

Imprint
Any brand names and product names mentioned in this book are subject to trademark, brand or patent protection and are trademarks or registered trademarks of their respective holders. The use of brand names, product names, common names, trade names, product descriptions etc. even without a particular marking in this work is in no way to be construed to mean that such names may be regarded as unrestricted in respect of trademark and brand protection legislation and could thus be used by anyone.

Cover image: www.ingimage.com

This book is a translation from the original published under ISBN 978-613-9-62857-5.

Publisher:
Sciencia Scripts
is a trademark of
Dodo Books Indian Ocean Ltd. and OmniScriptum S.R.L publishing group

120 High Road, East Finchley, London, N2 9ED, United Kingdom
Str. Armeneasca 28/1, office 1, Chisinau MD-2012, Republic of Moldova, Europe
Printed at: see last page
ISBN: 978-620-7-75285-0

CONTENTS

1. INTRODUCTION

Environmental pollution is one of today's main problems and is largely the responsibility of industrial activities, generating a large amount of waste, which is generally disposed of in the environment without efficient treatment.

In addition to visual pollution, the compounds are highly toxic and are considered to be carcinogenic, mutagenic and bioaccumulative in biota.

In this context, the pollution of surface water and groundwater is a major concern, since industrial, domestic and agricultural effluents are discharged directly into them.

Textile industrial activities consume a lot of water in their processes, generating a high volume of effluents and consequently contributing to the increase in contaminant levels in natural waters (SALLES et al., 2006).

Brazil, like India and China, has a high level of textile activity and is an important industrial sector for these countries.

The total production of dyes in the world is estimated at 800,000 tons per year (GANODERMAIERI *et al.*, 2005). In Brazil alone, it is estimated that more than 20,000 tons of these products are consumed annually by the textile industry (DALLAGO *et al.*, 2005).

The organic load released alters the ecosystem due to increased turbidity in the water, hinders the penetration of solar radiation, which generates changes in photosynthetic activity and the solubility regime of gases (SALLES *et al.*, 2006).

This fact, combined with the low use of inputs, especially dyes, means that the textile industry is responsible for generating waste with a high organic load and strong coloration (KUNZ *et al.,* 2002).

Around 10% of all the dye produced and used in the textile industry ends up being discharged into water, which can cause various environmental problems (GHAZI MOKRI et al., 2015).

Once in the environment, dyes and their derivatives can have mutagenic and carcinogenic effects on organisms (TOOR et al., 2006).

Once discarded into the environment, dyes prevent sunlight from penetrating the water, altering photosynthetic activity and decreasing oxygen solubility, thus damaging the entire aquatic environment (LALNUNHLIMI & KRISHNASWAMY, 2016). Dyes can also affect food and cause skin allergies and dermatitis (GHAZI MOKRI et al., 2015).

Many studies have therefore been carried out with the aim of proposing different methods of treating these textile dyes, which are discarded as waste by these industries.

Among these treatments are physicochemical methods such as ozonation (SANTOS et al., 2011), subcritical flow (HOSSEINI et al., 2010), flocculation (FURLAN et al., 2010) and photodegradation (BEHNAJADY et al., 2006). However, the high costs of these types of treatment emphasize the need for alternative methods (DILARRI et al., 2016a).

Biological treatment involving dye-degrading microorganisms has been studied and proposed as an alternative (PRIYA et al., 2015). However, degradation often requires a long period of time before achieving satisfactory results and can generate products that are more toxic than the initial dye itself.

Another problem is that the bacteria and fungi that degrade the dye are often opportunistic pathogens, such as *Candida albicans* (VITOR & CORSO 2008) and *Pseudomonas aeruginosa* (JADHAV et al., 2010).

One method of treating these textile dyes that has been studied is adsorption, the main characteristics of which are its low application cost, rapid waste containment time and the fact that it does not require a large physical area for treatment.

Another advantage of adsorption is the possibility of using an existing waste as an adsorbent material.

Various residues have already been tested as possible adsorbents, such as food peelings, agricultural waste, industrial waste such as sawdust, ashes, among others (SHARMA et al., 2011).

However, the availability of adsorbent material is often not sufficient to supply the volume of effluent that must be treated in the textile industry (DILARRI et al., 2016).

Another problem is that some materials, such as activated carbon, end up costing more, making it unfeasible to treat these effluents.

The yeast *Saccharomyces cerevisiae* is the most important microorganism in Brazilian industry, actively participating in various industrial processes (DILARRI et al., 2016a).

In addition, *S. cerevisiae* is produced on an industrial scale in Brazil, with tons of yeasts being made annually from this microorganism, which has a low market cost, as well as being a microorganism that is harmless to human health.

Another interesting factor of this microorganism is its cell wall, which has possible binding sites with metals and dyes, making it a possible adsorbent material for these residues.

Chitosan is a biopolymer synthesized from the shells of shrimp, crabs, lobsters and crayfish, which are some of the main fishing waste in Brazil (DILARRI et al., 2016b).

Chitosan has a low cost on the market, since it is currently produced on an industrial scale, with applications in various areas.

One of its potential applications is as an adsorbent material, capable of adsorbing various contaminants such as metals, dyes and organic residues.

Therefore, this work aimed to study these two materials in order to analyze all the various physical and chemical aspects of the adsorption of dyes by these compounds, being able to confirm which of the compounds would be economically and efficiently more advantageous for use in the treatment of textile industrial waste, thus improving the application of these materials as adsorbents.

2. OBJECTIVES

1.1 General objectives

- Adsorption of simulated effluent containing an isolated textile dye, using *S. cerevisiae* and chitosan powder as adsorbent material, and comparing the two types of adsorption, seeking to analyze the adsorbate/adsorbent ratio in both materials.

1.2 Specific objectives

• Improve the adsorption process in order to increase the sorption efficiency of the dye by *S. cerevisiae* and chitosan powder;

• Determine the possible binding sites between the dye, *S, cerevisiae*, and the chitosan powder, in order to confirm the type of interaction that will occur during the adsorption process.

3. MATERIAL AND METHODS

3.1 Chitosan

The chitosan powder, Figure 1, used in this work was purchased from the company Purifarma Quimica e Farmacèutica Ltda. Brazil-SP. This chitosan is derived only from shrimp shells and has a 90% degree of deacetylation, which corresponds to 5.60 x $10\text{-}3$ moles of amino group per gram of biopolymer. This chitosan powder has a density of 0.38 g mL^{-1} and a particle size of 80 mesh.

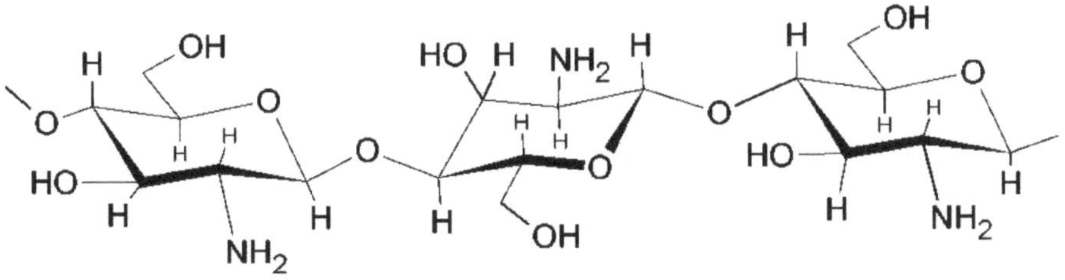

Figure 1 - Chemical structure of chitosan.

3.2 Yeast

The freeze-dried industrial strain of *S. cerevisiae*, obtained from the company Pak Gida Uretim ve Pazarlama A. S., batch number: 357723-30.

1.3 Dye

The dye used in this work was the azo dye Acid Blue 161 (AB 161), Figure 2, a soluble acid dye with a natural pH of 4.50. It was purchased from Aldrich Chemical Company Inc., under Chemical Abstract Service (CAS) number 12392-64-8, with a purity of 40% and a molecular mass of 416.39 g mol^{-1} .

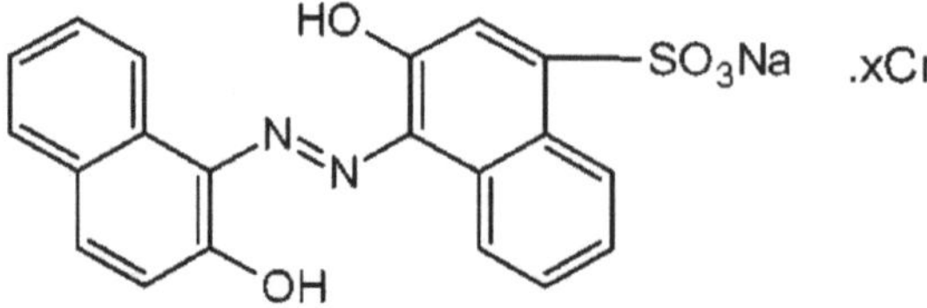

Figure 2 - Chemical structure of the Acid Blue 161 dye.

For the dye tests, a stock solution was prepared by diluting 1 g of the dye powder in 1 L of deionized water. After being homogenized, this solution was placed in an amber glass and stored in the dark.

1.4 Dye stability test

The stability of the dye was tested through its visible spectrum as the pH value of the solution varied. Samples of dye solution with a concentration of 100 μg mL^{-1} at different pH values (2.50, 4.50, 6.50 and 8.50) were analyzed in a Shimadzu Model 2401 - PC UV-Vis spectrophotometer, with scans between 200-800 nm, in a quartz cuvette with an optical path of 10 mm.

3.5 Study of adsorption kinetics

Kinetic studies were carried out using 20 mL of the dye solution at pH 4.50 with a concentration of 100 μg mL^{-1} in a 100 mL Erlenmeyer flask. A dry mass of 50 mg of the adsorbent materials tested was weighed and then these materials were placed in contact with the dye solution for 210 minutes at 293.15 K, with constant stirring at 40 rpm. Aliquots of the solution were removed every 30 minutes for analysis in a UV-Vis spectrophotometer, scanning from 200 to 800 nm. Before analyzing the samples in the spectrophotometer, the solutions were centrifuged for 20 minutes at 4000 rpm to separate the adsorbent material from the supernatant. Equation 1 was used to calculate the amount of dye adsorbed by each material.

$$qe = \frac{V.(Co - Ce)}{W}$$

(1)

where qe is the amount of dye adsorbed by the adsorbent material (μg mg^{-1}), Co is the initial concentration of the dye (μg mL^{-1}), Ce is the final concentration of the remaining dye (μg mL^{-1}), V is the volume of the solution (mL) and W is the mass of the adsorbent (mg).

For a better interpretation of the data, we used the mathematical models of pseudo-first order Lagergren (1898), Equation 2, and pseudo-second order Ho & McKay (1998) Equation 3.

$$\ln(qe - qt) = \ln qe - kl.t$$

(2)

$$\frac{t}{qt} = \frac{1}{ks.qe^2} + \frac{t}{qe}$$

(3)

where t is time (min), qt is the amount of dye adsorbed by the adsorbent in a given time t (μg mg$^-$ 1), kl is the pseudo-first order adsorption rate constant (min^{-1}) and ks is the pseudo-second order rate constant (mg μg^{-1} min).$^{-1}$

Kinetic processes are often controlled by diffusion, where the adsorbate is transferred to the inside of the adsorbent, thus penetrating beyond the outermost layer during the adsorption process. This process is known as intraparticle diffusion (ARAVINDHAN et al., 2007).

By confirming the occurrence of diffusion, it is possible to predict certain interactions between the adsorbate and the adsorbent.

The equation proposed by Webber & Morris (1963), Equation 4, was used to determine the occurrence of the diffusion process.

$$qt = Kdi.t^{0,5} + C \tag{4}$$

where Kdi is the intraparticle diffusion rate constant (mg µg^{-1} (min)$^{0,5-1}$), and C is the diffusion layer thickness constant (µg mg)$.^{-1}$

Intraparticle diffusion can occur simultaneously with surface adsorption and may not be the controlling factor in sorption (SABER-SAMANDARI & HEYDARIPOUR 2015).

To estimate the occurrence of this phenomenon, Equation 5 developed by Boyd et al. (1947) was applied.

$$F = \frac{qt}{qe}$$

e $\tag{5}$

$$Bt = -0,4977 - \ln(1 - F)$$

where F is the fraction of solute adsorbed at time t, and Bt is a mathematical function of F.

A graph of Bt vs t is made, and if the regression of this graph is linear with a passage through the origin, it means that intraparticle diffusion is the dominant adsorption process (ARAVINDHAN et al., 2007).

In this way, Boyd's second mathematical model, Equation 6, was used to calculate the intraparticle diffusion coefficients.

$$Di = \frac{r^2 . Bb}{\pi^2} \tag{6}$$

where Bb is calculated from the graph of Bt as a function of t, Di is the effective diffusion coefficient ($cm^2 \ s^{-1}$) of the dye in the adsorbents, and r is the radius of the adsorbent materials.

3.6 Study of isotherms

In order to analyze the different sorption mechanisms, tests were carried out using a 100 mL Erlenmeyer flask, with a 20 mL dye solution and a concentration of 100 µg mL^{-1}.

The pH of this solution was varied (2.50; 4.50; 6.50 and 8.50), as was the mass of the adsorbents, from 10 á 60 mg. The flasks were kept at a constant temperature of 293.15 K and stirred at 40 rpm.

The adsorbent materials were left in contact with the dye solution for a period of 4 hours before analysis in the UV-Vis spectrophotometer. All samples were centrifuged á 4000 rpm for 20 minutes before each analysis in the spectrophotometer.

For a better interpretation of the data obtained, the Langmuir and Freundlich models were applied.

The use of Langmuir and Freundlich isotherms helps to determine whether the adsorption mechanisms are chemical or physical, as well as indicating the existence of a strong or weak interaction between the adsorbent and adsorbate binding sites (PEARCE et al., 2003).

The Langmuir model is based on the fact that adsorption occurs on the surface of the adsorbent, where only specific sites will interact with the adsorbate, thus occupying a finite number of sites and forming monolayers (PODKOSCIELNY & NIESZPOREK, 2011).

Equation 7 shows Langmuir's mathematical model (1918).

$$\frac{Ce}{qe} = \frac{1}{l.qm} + \frac{Ce}{qm} \tag{7}$$

where l is the affinity between the adsorbent and the adsorbate (mL mg^{-1}), and qm is the maximum amount of dye adsorbed (μg mg).$^{-1}$

Using the graph of Ce/qe vs. Ce, it is possible to find the values of l and qm. Using Equation 8, proposed by McKay et al. (1982), it is possible to calculate the adsorption separation constant, defined as Rl.

In this mathematical model, if $Rl = 0$ the adsorption is irreversible, $Rl = 1$ the adsorption is linear, $Rl > 1$ is an unfavorable adsorption, and when $0 < Rl < 1$ indicates that the adsorption is favorable.

$$Rl = \frac{1}{1 + l.Co} \tag{8}$$

The Freundlich model (1906), shown in Equation 9, proposes that the adsorbent's binding sites are occupied exponentially, forming multiple layers (SABER-

SAMANDARI & HEYDARIPOUR, 2015).

This also indicates the heterogeneity of the surface of the adsorbent material, where more than one binding site will interact with the adsorbate, occupying these binding sites in decreasing order according to the intensity of the interaction between adsorbate and adsorbent.

$$\ln qe = \ln Kf + \frac{1}{nf}.\ln Ce \tag{9}$$

where nf is the Freundlich adsorption constant, and Kf is the sorption capacity constant of the solid (mL mg^{-1}). By plotting the linear regression of $lnqe$ vs. $lnCe$, it is possible to determine the values of nf and Kf.

3.7 Thermodynamic studies

Thermodynamic studies were carried out to assess whether adsorption is influenced by temperature.

Another advantage of thermodynamic studies is being able to assess whether adsorption is a spontaneous reaction, and whether it is an endothermic or exothermic process.

Thermodynamic studies were carried out by varying the temperature from 283.15 á 323.15 K.

A dye solution was used with a concentration of 100 µg mL^{-1} at pH 4.50 using a fixed amount of adsorbent mass of 50 mg.

Equation 10, proposed by van't Hoff, was used to find the values of the thermodynamic constants.

$$\ln Kts = \frac{\Delta s}{Rg} - \frac{\Delta H}{Rg} \cdot \frac{1}{T}$$

e

$$Kts = \frac{qe}{Ce}$$

(10)

where Rg *is* the universal gas constant (8.314 J mol^{-1} K^{-1}), T *is the* temperature (K), Kts is the equilibrium constant during temperature variation (mol g^{-1}), $-^2$ - is the entropy and is the enthalpy.

Using the *lnKts* vs. (1/T) graph, it is possible to determine the entropy and enthalpy values, and consequently, using Equation 11, it is possible to determine the Gibbs free energy ΔG values.

$$\Delta G = \Delta H - (\Delta S.T)$$

(11)

All the adsorption tests were carried out in triplicates, and to better confirm the data obtained, the standard deviation (*SD*) mathematical model, Equation 12, was used for the kinetic, isotherm and thermodynamic analyses. The lower the *SD* values, the more accurate the data estimate, as described by Gholizadeh et al. (2013).

$$DP = \sqrt{\frac{1}{N-1}\sum_{i=1}^{N}\left(\frac{Qie - Qic}{Qie}\right)^2}$$

(12)

Qie and *Qic* are the experimental values calculated from the dye adsorbed per mass of adsorbent material (μg mg^{-1}), and N is the number of experimental samples taken.

3.8 FT-IR spectrophotometer analysis

The adsorbent materials and the dye were analyzed in a Fourier Transform Infrared (FT-IR) spectrophotometer (Shimadzu 8300 model) before and after adsorption, with the aim of confirming the possible adsorbate/adsorbent interactions, identifying the main bonding sites, as well as being able to confirm whether adsorption occurs by a physical or chemical process.

To carry out these analyses, pellet salts were prepared using 1 mg of the dry sample and 149 mg of KBr.

The samples were then dried á 378.15 K for 24 hours and then pressed á 40 kN for 5 minutes, forming pellets for analysis. Thirty-two scans were carried out on each pellet using an FT-IR spectrophotometer with a range of 400 á 4000 cm^{-1} , with a resolution of 4 cm^{-1} .

All the graphs, chemical figures and calculations carried out in this work were made using Origin 6.10, ACD/ChemSketch and SciDAVis 1.D005 software.

4. Results and discussion

4.1 Dye stability tests

The dye showed no change in its spectrum as the pH changed, and can be considered a stable dye, showing no isobestic point or batochromic or hypsochromic effect.

However, the dye showed three chromophore groups, one in the range of 563.55 nm, a smaller one in the range of 373.92 nm and a maximum in the range of 603.82 nm, as shown in Figure 3.

Thus, the analysis of UV-vis spectrophotometry data can be interpreted by one of these three chromophore points.

Certain reactive functional groups can absorb different wavelengths in the visible spectrum, so they have more than one chromophore point (MITTER & CORSO, 2013).

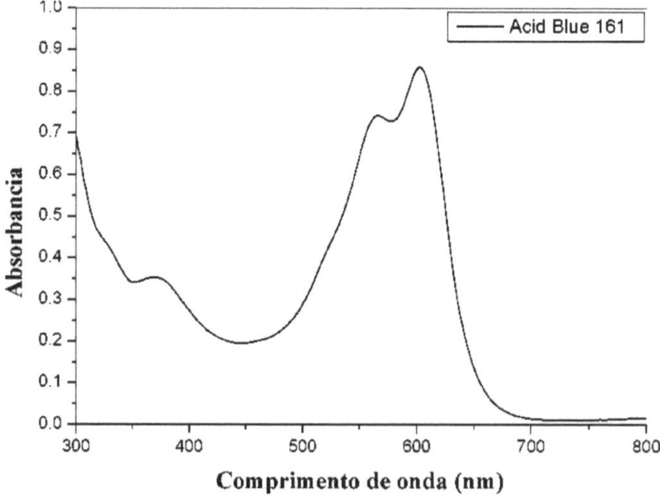

Figure 3 - Visible spectrum of the dye Acid Blue 161, at a concentration of 100 μg ml.$^{-1}$.

4.2 Study of adsorption kinetics

The kinetic studies showed that the $qe_{(exp)}$ of the chitosan powder reached equilibrium in 180 minutes, as shown in Figure 4, *while* the yeast cells reached equilibrium in 150 minutes, showing a faster sorption equilibrium than the chitosan powder.

In an industrial process, the retention time of the effluent is an important property, and the shorter the retention period, the more efficient the application of the technique (KUAI et al., 1999).

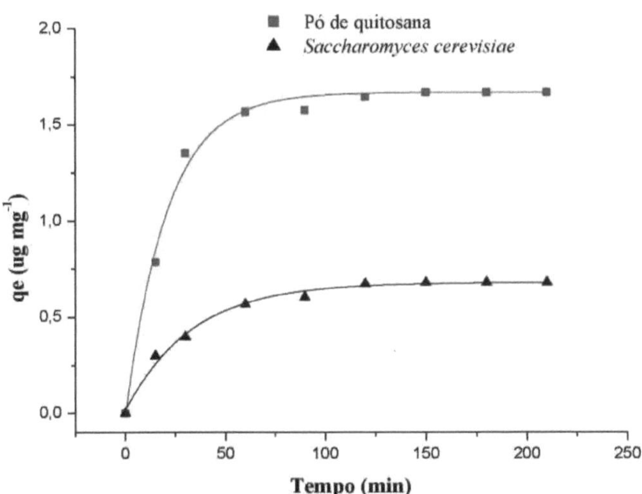

Figure 4 - Graph of the kinetics of dye adsorption by yeast and chitosan powder (Acid Blue 161 with a concentration of 100 μg mL^{-1} , 50 mg of adsorbent mass, 293.15 K temperature, 40 rpm stirring, total solution volume of 20 mL and a maximum contact time of 210 minutes).

Despite the shorter equilibrium time of *S, cerevisiae, the* $qe_{(exp)}$ value of the chitosan powder was much higher than that of the yeast, indicating that the chitosan powder has a greater capacity to adsorb the dye.

By making a linear regression of the Pseudo-first order graphs (*ln(qe-qt)* vs *t*) shown in Figure 5, and the Pseudo-second order (*(t/qt)* vs *t*) in Figure 6, it is possible to find the values of the constants of both mathematical models and establish which of the two kinetic models the adsorption is following.

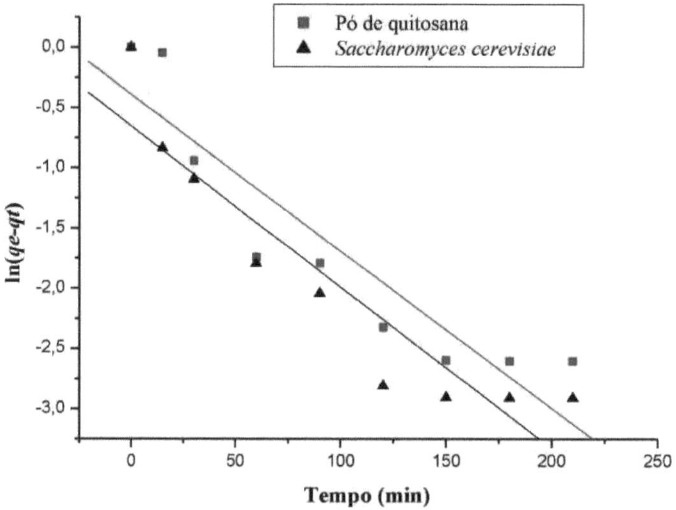

Figure 5 - Linear regression graph of the pseudo-first order model (Acid Blue 161 with a concentration of 100 µg mL-¹ , 50 mg of adsorbent mass, 293.15 K temperature, 40 rpm stirring, total solution volume of 20 mL and a maximum contact time of 210 minutes).

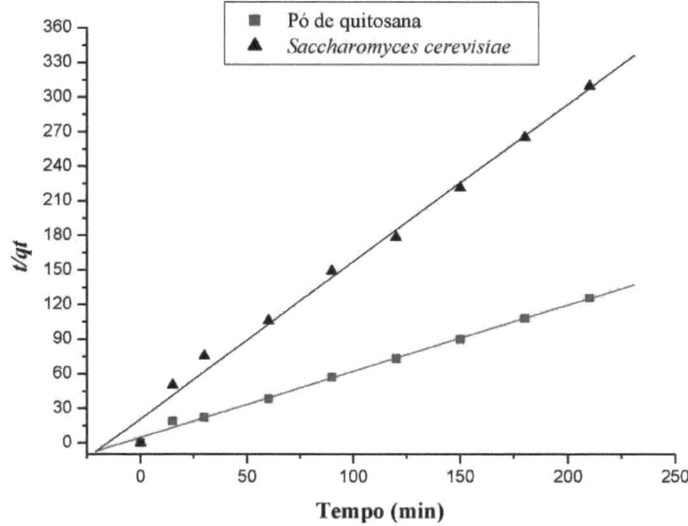

Figure 6 - Linear regression graph of the pseudo-second order model (Acid Blue 161 with a concentration of 100 µg mL^{-1} , 50 mg adsorbent mass, 293.15 K temperature, 40 rpm stirring, total solution volume of 20 mL and maximum contact time of 210 minutes).

Despite the difference in adsorbed dye capacity, both adsorbent materials best fitted the Pseudo-second order kinetic model.

This can be confirmed by the values of the correlation coefficient being higher in the Pseudo-second order model, and by the values of $q_{e(exp)}$ being closer to the values of $q_{e(cal)}$ in the Pseudo-second order model, as shown in Table 1.

Table 1. Results of kinetic studies

Pseudo first order	$q_{e(exp)}$ (µg mg)$^{-1}$	kl (min)$^{-1}$	q_{e} (cal) (µg mg)$^{-1}$	R^2	DP
S. cerevisiae	1,248	0,010	0,820	0,827	0,570
Chitosan powder	1,667	0,013	0,677	0,922	0,437

Pseudo second order	$qe_{(exp)}$ (µg mg^{-1})	ks (mg µg^{-1} min)$^{-1}$	$qe_{(cal)}$ (µg mg)$^{-1}$	R^2	DP
S. cerevisiae	1,248	0,034	1,407	0,982	5,722
Chitosan powder	1,667	0,069	1,741	0,998	2,886

When the adsorption process follows the Pseudo-second order model, it indicates that adsorption occurs through a process of chemical interaction by sharing or exchanging electrons between adsorbate and adsorbent (NANDI et al., 2009).

By plotting qt vs. $t^{0,5}$ it is possible to see two Iinear regions in both adsorbent materials; these regions represent the penetration of the dye into the adsorbent. The first layer is the surface of the material and the second layer is already diffused, as shown in Figures 7 and 8.

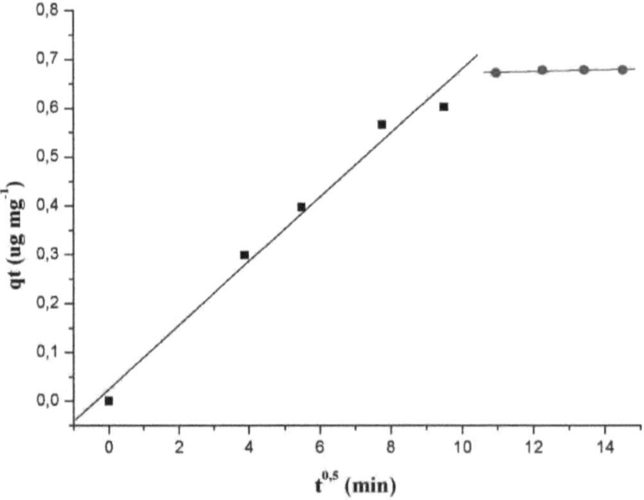

Figura 7 - Graph of intraparticle diffusion in *S. cerevisiae* (Acid Blue 161 with a concentration of 100 µg mL^{-1} , 50 mg adsorbent mass, 293.15 K temperature, 40 rpm agitation, total

solution volume of 20 mL and maximum contact time of 210 minutes).

In this way, it is possible to confirm the occurrence of intraparticle diffusion, i.e. the dye penetrates the interior of the adsorbent material.

It can also be seen that, since only a second adsorption line is formed, diffusion occurs in the macro pore layer for both materials.

The intraparticle diffusion regions begin in the initial boundary layer, followed by the macro, meso and micro pore layers (HO & MCKAY, 1998).

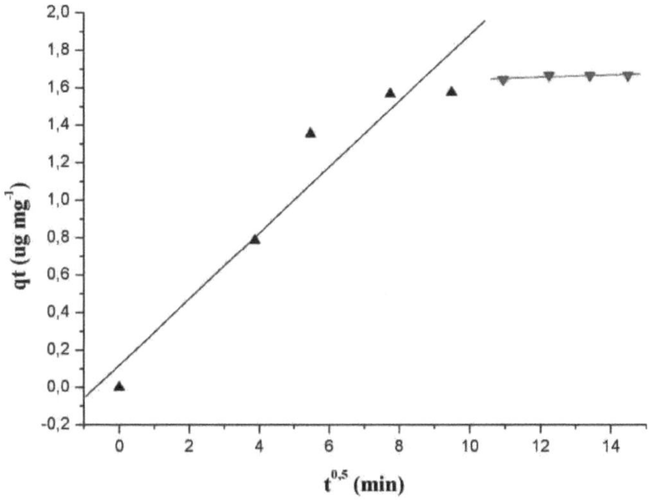

Figura 8   - Graph of intraparticle diffusion in chitosan powder (Acid Blue 161 with a concentration of 100 µg mL⁻¹ , 50 mg of adsorbent mass, 293.15 K temperature, 40 rpm stirring, total solution volume of 20 mL and a maximum contact time of 210 minutes).

The Boyd equation graph, Figure 9, showed that the linear regression line does not cross the origin in the adsorption of yeast and chitosan powder, so diffusion is not the controlling process of sorption in these materials.

The main binding sites for the dye are probably present on the surfaces of the materials, so surface sorption is the controlling process in adsorption.

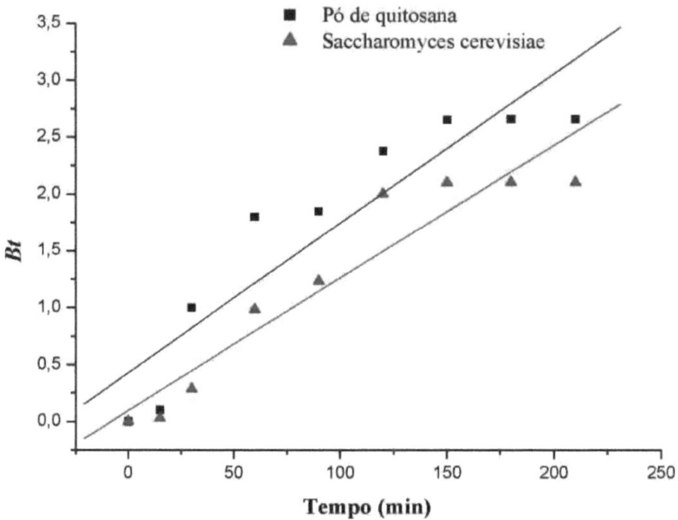

Figure 9 - Boyd function graph (Acid Blue 161 with a concentration of 100 µg mL^{-1} , 50 mg adsorbent mass, 293.15 K temperature, 40 rpm stirring, total solution volume of 20 mL and a maximum contact time of 210 minutes).

This result is confirmed by the *Di* values obtained, where chitosan powder obtained values of 1.043 x 10^{-7} cm^2 s^{-1} , and yeast 7.383 x 10^{-10} cm^2 s^{-1} , indicating that most of the interaction between the dye and the adsorbents occurs on their surfaces.

When the *Di* values remain in the order of 10^{-11} cm^2 s^{-1} this shows that the adsorption process is controlled by intraparticle diffusion, but when the *Di* values are higher than 10^{-11} cm^2 s^{-1} it means that adsorption is probably occurring for the most part only on the surface of the adsorbent (SINGH et al., 2003).

4.3 Study of isotherms

The results of the isotherms showed that regardless of the pH value of the solution, adsorption using the materials tested best fitted the Freundlich model, as shown in

Table 2.

Table 2. Results of the isotherms varying the pH values of the dye solution.

pH Yeast	Langmuir isotherm					Freundlich isotherm			
	Rl	qm	l	R^2	SD	Kf	nf	R^2	DP
2,5	0,171	1,299	0,044	0,932	0,639	$3,3 \times 10^{-4}$	0,301	0,982	0,186
4,5	0,444	0,161	0,011	0,979	5,082	$8,3 \times 10^{-5}$	0,534	0,981	0,143
6,5	0,540	0,106	0,007	0,858	2,530	8×10^{-12}	0,171	0,923	0,138
8,5	0,400	0,059	0,013	0,981	6,705	$1,3 \times 10^{-7}$	0,284	0,992	0,148

pH Chitosan	Langmuir isotherm					Freundlich isotherm			
	Rl	qm	l	R^2	SD	Kf	nf	R^2	DP
2,5	0,395	10,30	0,013	0,800	0,380	0,187	1,022	0,996	0,128
4,5	0,206	3,525	0,035	0,955	0,534	0,219	1,589	0,978	0,149
6,5	0,378	1,909	0,015	0,950	1,074	$9,1 \times 10^{-4}$	0,457	0,974	0,189
8,5	0,412	0,806	0,013	0,917	3,398	$1,1 \times 10^{-5}$	0,330	0,957	0,113

The Freundlich isotherm indicates that during the adsorption process more than one site of the adsorbent may be interacting with the dye, and that there is no connection between specific sites.

This result is corroborated by the heterogeneous composition of the yeast cell wall and the chitosan powder. The cell wall of *S. cerevisiae* is composed of chitin, mannoproteins and glucans (LIPKE & OVALLE, 1998).

Chitosan powder has a large number of monomers, with various bonding points such as OH and NH2 groups (DILARRI et al., 2016b).

The Freundlich isotherm also confirms that adsorption on these two adsorbents takes place in multiple layers, with lateral interactions occurring during adsorption.

Acidic pH probably influences the increase in bonding layers, since it is at this pH that the values obtained for *Kf* and *nf* were highest.

Although adsorption does not follow the Langmuir model, it can be seen that the *qm* values of chitosan powder and yeast are higher at the more acidic pHs, which are 2.50 and 4.50.

This result confirms that adsorption is more efficient at an acidic pH, increasing the interaction between the dye and the adsorbent material. The acidic pH of the dye solution can lead to a protonation of the binding sites of the adsorbent material, as well as increasing the interaction by hydrogen bridges in the solution, facilitating adsorption (MENDES et al., 2015).

The *qm* values of the chitosan powder were higher than those of the yeast in all the experiments, thus confirming its greater ability to adsorb the dye from the solution.

All the *Rl* values obtained showed that adsorption was favorable, even when changing the pH of the solution. Although sorption was lower at pH 6.50 and 8.50, it is possible to state that both adsorbents showed affinity with the dye even at these pH values.

4.4 Thermodynamic studies

Thermodynamic studies have shown that both adsorbents are influenced by temperature, i.e. the higher the temperature of the solution, the more efficient the adsorption.

The chitosan powder and *S. cerevisiae* did not show any change or deformation during the temperature increase, but this is reported by Suteu et al.

(2013), the biomass of *S, cerevisiae* tends to deteriorate at ȧ 318.15 K, greatly reducing its sorption capacity.

However, this phenomenon was not observed in this study. This divergence is probably due to the use of different strains in each of the studies, consequently presenting different results (DILARRI et al., 2016a).

The positive enthalpy values, as shown in Table 3, confirm that adsorption is an endothermic process for both adsorbent materials tested.

Positive entropy values indicate randomness in the bonds at the solid-solution interface, showing that more than one bonding site is acting on the adsorbent (DILARRI et al., 2016a). This is also indicative of interaction with heterogeneous sites, corroborating the data obtained from the isotherms.

Table 3 Thermodynamic constant values of the adsorbents tested

	ΔS	ΔH	ΔG			
	(kJ mol^{-1} K^{-1})(kJ mol)$^{-1}$		(kJ mol)$^{-1}$			
			283,15 K	293,15 K	303,15 K	323,15 K
Chitosan	0,0015	7,461	7,036	7,021	7,006	6,976
Yeast	0,065	31,360	12,955	12,305	11,655	10,355

Looking at the Gibbs free energy values, it is possible to see that it decreases as the temperature increases, which confirms that the adsorption between the AB 161 dye and the chitosan powder and *S, cerevisiae* is a spontaneous reaction. This also confirms that the dye has a high affinity for chitosan powder and *S, cerevisiae.*

4.5 FT-IR spectrophotometer analysis

The FT-IR spectrum of the dye, shown in Figure 10, showed intense peaks in the 1369, 1141 and 667 cm regions^{-1} , which indicate vibrations of -C-H, -C=C-, and =C-C bonds in aromatic rings. The band at 1566 cm^{-1} indicates the vibration of the -N=N- azo bond present in the dye molecule. The 1598 cm band^{-1} is the stretching of the azo bond with a -CH group.

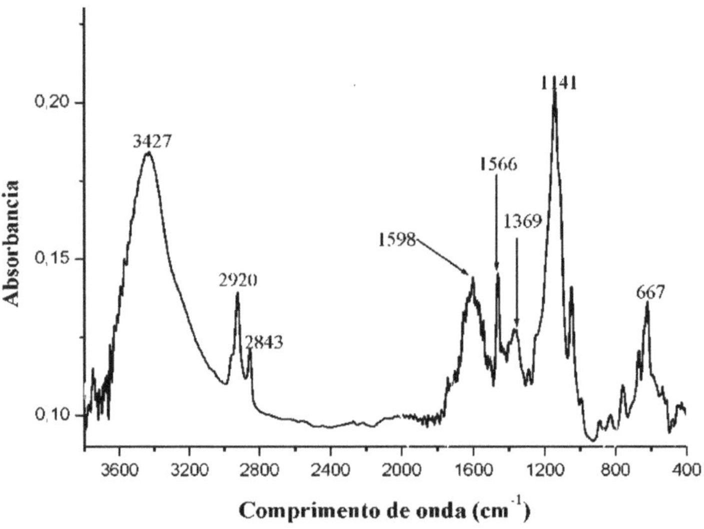

Figure 10 - Graph of the FT-IR spectrum of the Acid Blue 161 dye.

This spectrum corroborates the chemical molecule of the dye.

The bands in the region 3427, 2920 and 2843 cm^{-1} are vibrations of amine groups and carbon residues from the air.

Looking at the FT-IR spectrum of chitosan powder, shown in Figure 11, it is possible to see a high intensity in the 1049 cm band^{-1} , which indicates the vibration of -C-O groups present in the biopolymer monomers.

Other intense bands are in the region 1659 and 1377 cm^{-1} , which are bonds with the chemical groups $-NH_2$ and $-CH-NH_2$.

Another band in this spectrum that stands out is in the 623 cm region^{-1} , which is the stretching of the -C-OOH bond.

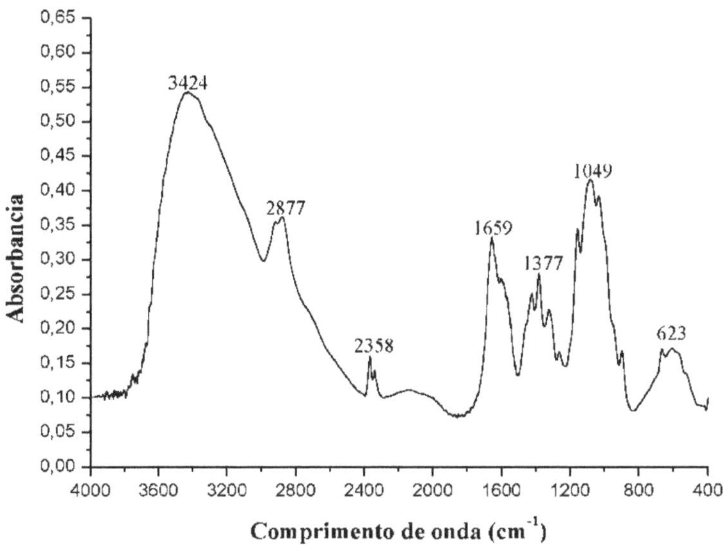

Figure 11 - Graph of the FT-IR spectrum of chitosan powder.

The bands at 3424 cm^{-1} are free primary amine groups. The bands in the region 2877 and 2358 cm^{-1} are residues of CO_2 present in the environment.

The spectrum of chitosan obtained in this work was very similar to others already reported by Ko et al. (2012) and Carneiro et al. (2014), thus corroborating the characterization carried out in this analysis.

The FT-IR spectrum of *S. cerevisiae*, shown in Figure 12, showed two intense bands in the region 1540 to 1649 cm^{-1} , this region is the vibration of the starch group R-NH-C-O-CH3, present in the chitin of the yeast cell wall.

The peak at 1350 cm^{-1} is the stretching of the -C-NH bond, which is also a chemical group belonging to the chitin molecule in the yeast cell wall.

The 1234 cm band^{-1} represents the vibration of the -C=O bond. The peak at 1039 cm^{-1} is the vibration of -C-O groups, typical of sugars present in the yeast cell wall (DILARRI et al., 2016a).

The band in the 558 cm region^{-1} resonates with the presence of -C-N-C groups found in yeast membrane and cell wall proteins (DILARRI et al., 2016).

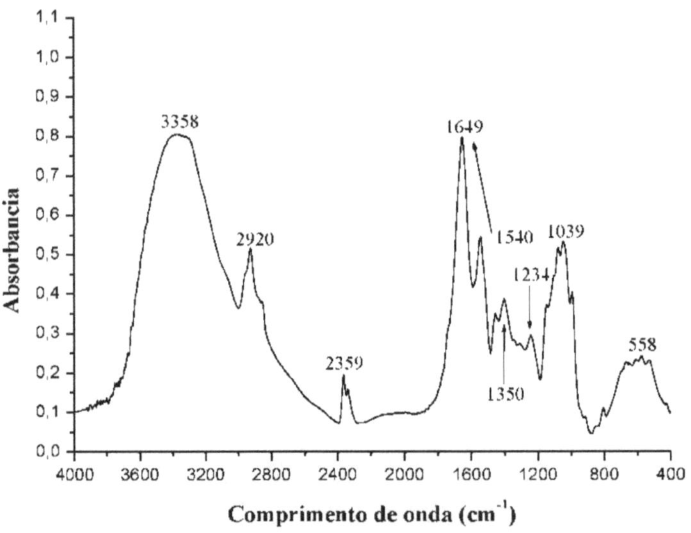

Figure 12 - Graph of the FT-IR spectrum of *S. cerevisiae* yeast.

The 2359 and 2920 cm bands^{-1} are residues of atmospheric carbon dioxide and free -CH groups.

The peak in the region 3358 cm^{-1} indicates free amine groups. What stands out in this characterization is the large presence of chemical groups related to the chitin in the yeast cell wall.

This presence of groups related to chitin, as well as groups present in sugars and proteins of the cell wall and membrane are also reported in the work carried out by Suteu et al. (2013).

When the FT-IR spectrophotometer analyses were carried out after adsorption at pH 2.50 and 8.50, certain changes in the spectra could be seen.

The FT-IR spectrophotometer is a very useful tool for identifying adsorption mechanisms and can confirm whether adsorption occurs through a chemical or physical process (MONASH & PUGAZHENTI, 2009).

Looking at the spectra of the chitosan powder after adsorption at both pHs, as shown in Figure 13, it is possible to see that the spectrum after adsorption at pH 8.50 shows no change when compared to the spectrum before adsorption.

Even the intensity of its bands is very similar to the spectrum before adsorption, which confirms the occurrence of physisorption.

The FT-IR spectrum is like a fingerprint of the molecule, its change indicates that a new chemical interaction or breakdown of the molecule has occurred (DILARRI et al., 2016a).

Thus, this lack of change in the chitosan powder molecule after adsorption at pH 8.50 confirms the results obtained from the isotherms, showing that at alkaline pHs adsorption is probably occurring by van der Waals forces.

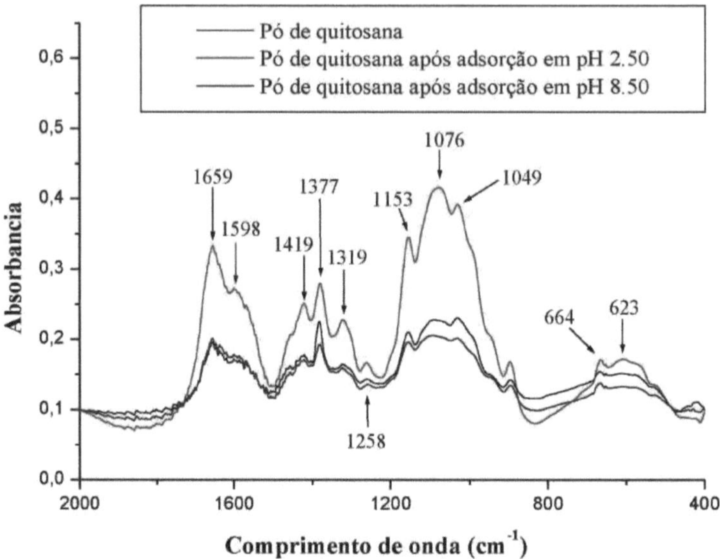

Figura 13 - Graph of the FT-IR spectrum of chitosan powder before and after adsorption of the AB 161 dye at pH 2.50 and 8.50.

Physical adsorption is characterized by low adsorbate/adsorbent affinity and lower interaction strength compared to chemical adsorption.

This result explains why the worst adsorption results were when the pH of the medium was 6.50 and 8.50.

When looking at the spectrum of the chitosan powder after adsorption in acidic pH, it is possible to see greater intensity in its peaks, as well as new bands and some deformations in certain regions.

The bands 1659 and 1377 cm^{-1} showed greater intensity, indicating that some interaction is taking place in this region, which is the vibration of the -C-NH2 chemical group.

The peak in the region 1598 cm^{-1} is a deformation, present only in the spectrum of chitosan powder after adsorption at pH 2.50, which represents the stretching of the molecule -NH3^{+} , which is a protonated group present in chitosan monomers. One of the advantages of adsorption at acidic pH is the possible protonation of certain interaction sites of the adsorbent, increasing sorption efficiency (MENDES et al., 2015).

The bands 1419 and 1319 cm^{-1} also showed high intensity. The 1153 and 1076 cm peaks^{-1} are singular deformations present only in the FT-IR spectrum of the chitosan powder after adsorption at pH 2.50. These bands are the vibrations of sulfonic groups present in the dye molecule, which are probably interacting by chemical bonds with the chitosan powder.

The 1049 cm band^{-1} corresponds to bonds between -C-C groups in cyclic chains or aromatic rings, probably belonging to the dye molecule.

The bands 1258, 664 and 623 cm^{-1} also showed slight deformations and greater intensity, indicating a possible interaction with the dye in these regions. Due to the H ions^{+} present in acidic solutions, it is likely that hydrogen bridge interactions are occurring, justifying the slight deformations and higher intensities in the bands (MENDES et al., 2015).

It is therefore possible to confirm that chemosorption is occurring at acidic pH, corroborating the data obtained in the isotherm studies.

By analyzing the FT-IR spectrum of the yeast before and after adsorption at pH 2.50 and 8.50, shown in Figure 14, it is possible to see that just like the chitosan powder, at alkaline pH there was no change in the spectrum.

Small increases in intensity can be observed, but even so the spectrum remained similar to the spectrum before adsorption, and it can be concluded that adsorption at alkaline pH occurs through a physical process.

This result is similar to the chitosan powder and is corroborated by the data obtained in the isotherm studies.

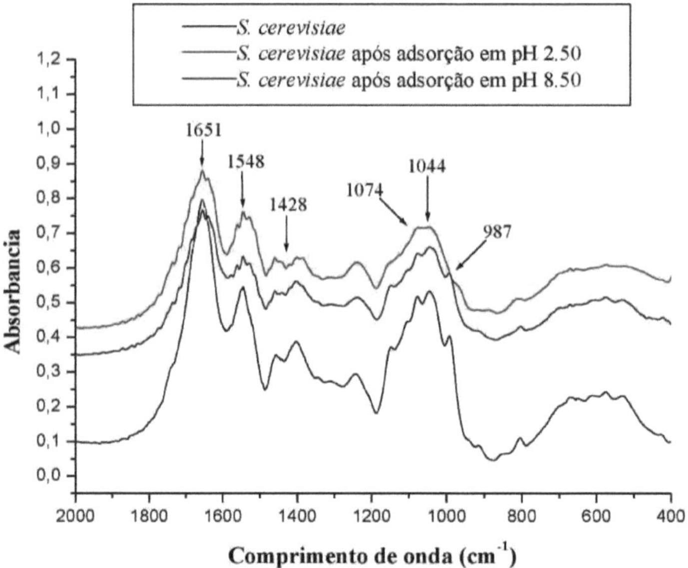

Figura 14 - Graph of the FT-IR spectrum of *S. cerevisiae* before and after adsorption of the AB 161 dye at pH 2.50 and 8.50.

The FT-IR spectrum of *S. cerevisiae* after adsorption at pH 2.50 showed certain singularities, as well as more intense bands.

The band in the region 1651 cm^{-1} kept the same pattern, being slightly more intense at pH 2.50.

It is possible to see some changes in the 1548 cm peak^{-1} , as well as a greater intensification. This region represents the stretching of the -N-H bond, indicating a possible interaction with the dye.

The band in the 1428 cm region^{-1} shows a distinct difference from the other spectra. This region represents the stretching of the -C-N bond, which is a probable binding site for the dye.

At acidic pH, amine and amide groups tend to become protonated, allowing them to interact chemically with other molecules present in the solution (MENDES et al., 2015).

This is probably a site of interaction between the yeast and the AB 161 dye.

The peaks in the 1074 and 1044 cm regions^{-1} also showed changes when compared to the other two spectra. This region is the vibration of the -R-C- $_{NH2}$ group, which probably belongs to the yeast cell wall chitin, and is probably acting as a binding site for the dye.

Chitin is probably the main site of adsorbate/adsorbent interaction with the dye. Sharma et al. (2011) reported on the potential for using fungal biomass to adsorb dyes, as its cell wall made up of chitin is the main binding site for the dye. It is therefore natural for chitin to stand out as a binding site.

Another change in the spectrum is seen in the 987 cm region^{-1} , this band is associated with sugars present in the yeast membrane and cell wall, which may be interacting with the dye, generating this deformation.

With these results obtained in the FT-IR spectrophotometer analysis, together with the data obtained in the isotherm studies, it is possible to state that at acidic pH, chemisorption is occurring between the yeast *S', cerevisiae* and the dye AB 161.

5. CONCLUSION

The dye did not undergo any changes in its spectrum as the pH values of the solution changed, thus proving to be stable.

The dye also had three chromophore points in its spectrum, allowing its removal to be analyzed by quantifying any of these three points, with the maximum wavelength used in this work being 603.82 nm.

The kinetic studies showed that adsorption reaches equilibrium more quickly in the yeast *S. cerevisiae*, but the chitosan powder showed a greater adsorptive capacity than the yeast.

The kinetic studies also confirmed that both materials reach equilibrium in less than 180 minutes, which is an important property in wastewater treatment.

The kinetic studies also showed that the two adsorptions best fit the Pseudo-second order kinetic model.

Intraparticle diffusion analysis showed that the dye penetrates the yeast and chitosan powder, but this is not the controlling process of sorption, so adsorption occurs mainly on the surface of the adsorbents.

The isotherm studies showed that the adsorption had a better fit to the Freundlich model, even changing the pH of the solutions, for both adsorbent materials.

It can be concluded that adsorption occurs through the formation of multiple layers and lateral interactions.

The results obtained from the Langmuir constants also indicated that adsorption was favorable at all pH levels, showing a strong interaction between the dye and the two adsorbent materials.

It was also possible to prove that chitosan powder has a greater adsorptive capacity, managing to remove the dye more efficiently from the solution than the *S. cerevisiae* yeast.

Thermodynamic studies confirmed that for both adsorbent materials adsorption is a spontaneous, endothermic reaction.

Thermodynamic studies have also shown that adsorption is influenced by temperature, with the higher the temperature the more efficient the adsorption for both adsorbents.

FT-IR spectrophotometer characterization showed the main chemical groups of the dye and the two adsorbents tested. FT-IR spectrophotometer analysis also showed that chitosan powder and *S. cerevisiae* are chemisorbing the dye at acidic pH, due to the changes and deformations in the spectra.

However, at alkaline pH it was concluded that a weaker interaction by physisorption was occurring, but there was no change in the spectra of the chitosan powder and *S. cerevisiae.*

The analyses also helped to confirm the results obtained in the isotherm studies. It can therefore be concluded that pH has a direct influence on adsorption, being more efficient at acidic pH.

Both adsorbents proved to be good adsorbent materials, but due to their greater sorption capacity, practicality of use and rapid settling, it is concluded that chitosan powder is a much more efficient and applicable adsorbent for treating effluents contaminated by textile dyes, making it a possible alternative for the bioremediation and treatment of these compounds.

6. BIBLIOGRAPHY

ARAVINDHAN, R.; RAO, J. R.; NAIR, B. U. Removal of basic yellow dye from aqueous solution by sorption on green algae *Caulerpa scalpelliformis*. **Journal of Hazardous Materials**, v. 142, p. 68-76, 2007.

BOYD, G. E.; ADAMSON, A. W.; MYERS, L. S. The exchange adsorption of ions from aqueous solution by organic zeolites, II. Kinetics. **Journal of the American Chemical Society**, v. 69, p. 2836-2848, 1947.

BEHNAJADY, M. A.; VAHID, B.; MODIRSHAHLA, N.; SHOKRI, M. Evaluation of electrical energy per order (EEO) with kinetic modeling on photooxidative degradation of C.I. Acid Orange 7 in a tubular continuous-flow photoreactor. **Industrial & Engineering Chemistry Research**, v. 45, p. 553-557, 2006.

CARNEIRO, L. A. B. C.; COSTA-SILVA, T. A.; SOUZA, C. R. F.; BACHAMANN, L.; OLIVEIRA, W. P.; SAID, S. Immobilization of lipases produced by the endophytic fungus *Cercospora kikuchiion on* chitosan microparticles. **Brazilian Archives of Biology and Technology**, v. 57, p. 578-586, 2014.

DALLAGO, R. M.; SMANIOTTO, A.; OLIVEIRA,L. C. A. Solid tannery waste as adsorbents for the removal of dyes in aqueous media. *Quimica Nova*, v. 28, p. 433-437, 2005.

DILARRI, G.; ALMEIDA, E. J. R.; PECORA, H. B.; CORSO, C. R. Removal of Dye Toxicity from an Aqueous Solution Using an Industrial Strain of *Saccharomyces cerevisiae* (Meyen). **Water, Air, & Soil Pollution**, v. 227, p. 269, 2016a.

DILARRI, G.; MENDES C. R.; MARTINS, A. O. Synthesis of chitosan biofilms crosslinked with tripolyphosphate acting as a chelating agent in the fixation of silver nanoparticles. **Ciência & Engenharia**, v. 25, n. 1, p. 97-103, 2016b.

FREUNDLICH, H. (1906). Adsorption in solution. **Zeitschrift für Physikalische Chemie**, v. 40, p. 1361-1368, 1906.

FURLAN, F. R.; SILVA, L. G.; MORGADO, A. F.; SOUZA, A. A.; SOUZA, S. M. Removal of reactive dyes from aqueous solution using combined coagulation/flocculation and adsorption on activated carbon. **Resource, Conservation and Recycling**, v. 54, p. 283-290, 2010.

GANODERMAIERI, G., CENNAMO, G., SANNIA, G., Remazol Brilliant Blue R decolourization by the fungus *Pleurotus ostreatus* and its oxidative enzymatic system. *Enzyme and Microbial Technology*, v. 36, p. 17-24, 2005.

GHAZI MOKRI, H. S.; MODIRSHAHLA, N.; BEHNAJADY, M. A.; VAHID, B. Adsorption of C.I. Acid Red 97 dye from aqueous solution onto walnut shell: kinetics, thermodynamics parameters, isotherms. **International Journal of Environmental Science and Technology**, v. 12, p. 1401-1408, 2015.

GHOLIZADEH, A.; KERMANI, M.; GHOLAMI, M.; FARZADKIA, M. Kinetic and isotherm studies of adsorption and biosorption processes in the removal of phenolic compounds from aqueous solutions: Comparative study. **Journal of Environmental Health Science & Engineering**, v. 11, p. 29, 2013.

HO, Y. S.; MCKAY, G. Sorption of dye from aqueous solution by peat. **Chemical Engineering Journal**, v. 70, p. 115-124, 1998.

HOSSEINI, S. D.; ASGHARI, F. S.; YOSHIDA, H. Decomposition and discoloration of synthetic dyes using hot/liquid (subcritical) water. **Water Research**, v. 44, p. 1900-1908, 2010.

JADHAV, J. P.; PHUGARE, S. S.; DHANVE, R. S.; JADHAV, S. B. Rapid biodegradation and decolorization of Direct Orange 39 (Orange TGLL) by an isolated bacterium *Pseudomonas aeruginosa* strain BCH. **Biodegradation**, v. 21, p. 453-463, 2010.

KUAI, L.; KERSTENS, W.; CUONG, N. P.; VERSTRAETE, W. Treatment of domestic wastewater by enhanced primary decantation and subsequent naturally ventilated trickling filtration. **Water, Air, & Soil Pollution**, v. 113, p. 43-62, 1999.

KUNZ, A.; PERALTA-ZAMORA, P.; MORAES, S. G.; DURAN, N. New trends in the treatment of textile effluents. *Quimica Nova,* v. *25,* n. 1, p.78-82, 2002.

KO, Y. G.; LEE, H. J.; SHIN, S. S.; CHOI, U. S. Dipolar-molecule complexed chitosan carboxylate, phosphate and sulphate dispersed electrorheological suspensions. **Soft Matter**, v. 23, p. 6273-6279, 2012.

LAGERGREN, S. Zur theorie der sogenannten adsorption gel oster stoffe. Kung-liga Svenska Vetenskapsakademiens. **Handlingar**, v. 24, p. 1-39, 1898.

LANGMUIR, I. The adsorption of gases on plane surface of glass, mica and platinum. **Journal of the American Chemical Society**, v. 40, p. 1361-1368, 1918.

LALNUNHLIMI, S.; KRISHNASWAMY, V. Decolorization of azo dyes (Direct Blue 151 and Direct Red 31) by moderately alkaliphilic bacterial consortium. **Brazilian Journal of Microbiology**, v. 47, p. 39-46, 2016.

LIPKE, P. N.; OVALLE, R. Cell wall architecture in yeast: New structure and new challenge. **Journal of Bacteriology**, v. 180, p. 3735-3740, 1998.

MCKAY, G.; BLAIR, H. S.; GARDNER, J. R. Adsorption of dyes on chitin: Equilibrium studies. **Journal of Applied Polymer Science**, v. 27, p. 3043-3043, 1982.

MENDES, C. R.; DILARRI, G.; PELEGRINI, R. T. Application of biomass *Saccharomyces cerevisiae* as adsorption agent of dye Direct Orange 2GL and possible interactions mechanisms adsorbate/adsorbent. **Materia (Rio de Janeiro)**, v. 20, p. 898-908, 2015.

MITTER, E. K.; CORSO, C. R. Acid dye biodegradation using *Saccharomyces cerevisiae* immobilized with polyethyleneimine-treated sugarcane bagasse. **Water, Air, & Soil Pollution**, v. 224, p. 1391-1398, 2013.

MONASH, P.; PUGAZHENTI, G. Adsorption of Crystal Violet dye from aqueous solution using mesoporous materials synthesized at room temperature. **Adsorption**, v. 15 p. 390-405, 2009.

NANDI, B. K.; GOSWAMI, A.; PURKAIT, M. K. Adsorption characteristics of brilliant green dye on kaolin. **Journal of Hazardous Materials**, v. 161, p. 387-395, 2009.

PEARCE, C. I.; LLOYD, J. R.; GUTHRIE, J. T. The removal of color from textile wastewater using whole bacterial cells: A review. **Dyes and Pigments**, v. 58, p. 179196, 2003.

PODKOSCIELNY, P.; NIESZPOREK, K. Adsorption of phenols from aqueous solutions: Equilibria, calorimetry and kinetics of adsorption. **Journal of Colloid and Interface Science**, v. 354, p. 282-291, 2011.

PRIYA, E. S.; SELVAN, P. S.; UMAYAL A. N. Biodegradation studies on dye effluents and selective Remazol dyes by indigenous bacterial species through spectral characterization. **Desalination and Water treatment**, v. 55, p. 241-251, 2015.

SABER-SAMANDARI, S.; HEYDARIPOUR, J. Onion membrane: An efficient adsorbent for decoloring of wastewater. **Journal of Environmental Health Science**

& Engineering, v. 13, p. 16, 2015.

SALLES, P. T. F.; PELEGRINI, N. N. B.; PELEGRINI, R. T. Electrochemical treatment of industrial effluent containing reactive dyes. **Engenharia Ambiental**, v. 3, n. 2, p. 25-40, 2006.

SANTOS, P. K.; FERNANDES, K. C.; FARIA, L. A.; FREITAS, A. C.; SILVA, L. M. Decolorization and degradation of the red azo dye GRLX-220 by ozonation. **Quimica Nova**, v. 34, n. 8, p. 1315-1322, 2011.

SHARMA, P.; KAUR, H.; SHARMA, M.; SAHORE, V. A review on applicability of naturally available adsorbents for the removal of hazardous dyes from aqueous waste. **Environmental Monitoring and Assessment**, v. 183, p. 151-195, 2011.

SINGH, K. P.; MOHAN, D.; SINHA, S.; TONDON, G. S.; GOSH, D. Color removal from wastewater using low-cost activated carbon derived from agricultural waste material. **Industrial & Engineering Chemistry Research**, v. 42, p. 1965-1976, 2003.

SUTEU, D.; BLAGA, A. C.; DIACONU, M.; MALUTAN, T. Biosorption of reactive dye from aqueous media using *Saccharomyces cerevisiae* biomass. Equilibrium and kinetic study. **Central European Journal of Chemistry**, v. 11, p. 2048-2057, 2013.

TOOR, A. P.; VERMA, A.; JOTSHI, C. K.; BAJPAI, P. K.; SINGH, V.

Photocatalytic degradation of direct yellow 12 dye using UV/TiO, in a shallow pond slurry reactor. **Dyes and Pigments**, v. 68, p. 53-60, 2006.

VITOR, V.; CORSO, C. R. Decolorization of textile dye by *Candida albicans* isolated from industrial effluents. **Journal of Industrial Microbiology and Biotechnology**, v. 35, p. 1353-1357, 2008.

WEBER, W. J.; MORRIS, J. C. Kinetics of adsorption on carbon from solution. **Journal of the Sanitary Engineering Division**, v. 89, p. 31-60, 1963.

Printed by Books on Demand GmbH, Norderstedt / Germany